195+ WAYS TO PROTECT YOUR PRIVACY AND CONFIDENTIALITY

SIMPLE & EASY TIPS FOR DIGITAL & FINANCIAL LITERACY

VIJAYKUMAR GUMMADI

ISBN 979-888546217-4

To My Mother

Mrs Baby Samrajyam Gummadi

Contents

Contents

Foreword

Hectic modern era caused wide spread digital and financial illiteracy due to lack of time & ignorance.

Main reason for propagation of this digital and financial illiteracy is people are reluctant as well as developed aversion for continuing their learning w.r.t. need of the time as per changing technologies.

Privacy and Confidentiality protection has become basic essential requirement to all in this modern world and can no longer be just ignored and it has become part of everyone life.

Being a Chemical Engineering & Business Management Professional associated with nuclear sector gave me deep insight on digital and financial illiteracy.

You can't always reach out to an expert or specialist for guidance. I have compiled a set of 195+ useful tips that can be helpful to protect yourself.

VIJAYKUMAR GUMMADI, Author

CHAPTER ONE

GENERAL

1

Make your privacy a priority.

2

Devices can be your computers, laptops, tabs, mobile phones etc

3

Technology is continuously evolving. Your device has hardware, operating system software, application software etc. Devices functional output is a result of perfect integration of above three aspects. Device can be never intelligent it is only logically mechanical. You can't rely or depend blindly on any device.

4

Due to technological leaps obsolescence of devices is at rapid phase resulting in frequent changes in hardware, operating system software, application software. Hence continuous & perfect integration among above aspects is difficult due to which there can be security & privacy breaches time to time. No device can be fool proof.

5

Humans are responsible for their own privacy. They must remain alert, attentive, and vigilant all the times especially while executing important activities using devices.

CHAPTER TWO

INFORMATION SHARING

6

Practice only mindful & conscious sharing.

7

Never share your sensitive number details of driving licence, credit card, bank details, ID proofs etc either in writing or through internet with anyone. If sharing of numbers is compulsory provide only last few digits of number.

8

Beware of FREE, FREE, FREE campaigns with request to fill forms with your details, either they may extract & sell your personal data or they may levy hidden charges

9

Medical records are confidential and should not be shared with others.

10

While sharing private information insist for receipt signature from specific recipient person. This is required for authenticity and to prevent misuse.

11

Don't respond to either online or offline surveys ads. Surveys look lucrative with easy gifts and monetary benefits. However they insist to fill your personal as well as professional details that may result in breach of your privacy.

12

While using photo copier check it's settings & ensure it's not creating any parallel scanned soft copies & post copying take back originals.

13

Never avail any promotional loyalty program such as discounts, money back, points etc which insists to fill your personal details and data

14

Birthday represents your identity, hence never share your birthday with anyone when not required.

15

Don't share your entertainment or other paid subscription accounts even with close friends as they contain your personal details including payment details

16

Never permit online auto filling of forms with your personal details. Be mindful and prevent internet from auto saving of your personal credentials

17

Never try to share photos online especially through social media platforms.

18

Be cautious while clicking any delivered text link. Verify credentials before clicking.

19

Password protect vital documents and shared files.

20

Avoid sharing files through chat platforms. Share individually through direct Email.

21

Hide your friends list.

22

Keep your photos and family albums private

23

Hide your personal information and credentials wherever possible.

24

When you had to share your personal information insist for safeguarding of your credential information from falling into third party.

CHAPTER THREE

LOGIN & ACCOUNTS

25

Access all your smart devices/web accounts with individual passwords with periodic password changing. Wherever possible use two factor security be authentication.

26

Use auto lock screen. Don't share phone unlock PIN other than your family

27

Always use strong passwords. Rhyming sentences along with alpha numeric characters is preferred secured password.

28

Never use face unlock. Face unlock may not be secure always.

29

Delete Emails /SMS received for new password generation during password recovery or reset.

30

Never save auto fill passwords while surfing in browsers

31

Change auto generated login passwords immediately.

32

Use security questions & other alternative methods for account recovery on forgot password

33

Never share login data with third party apps

CHAPTER FOUR

COMMUNICATION & CORRESPONDENCE

34

Never open or react to anonymous Emails or SMS. Just block those Email IDs and numbers permanently. They can intrude into your device and extract your information.

35

Don't respond to phone calls or Emails claiming on behalf of banks are financial institutions falsely seeking your personal details on the context of some direct or indirect benefits to you. They can be copycats stealing your identity. Verify their credentials with customer care helpline numbers or Email IDs. Block and bar those numbers and Email IDs.

36

Never send private documents to common Email ID or FaX or general address. Always share private documents to only relevant specific person credentials and also seek confirmation receipt. Otherwise your valuable documents may be lost or misplaced or misused.

37

On vacation use Auto reply Email and inform when you will back and incase of urgency to whom they have to contact.

38

Auto forward alternate Email/mobile call to main Email/phone wherever required.

39

Don't touch phishing and spurious Emails.

40

Delete unknown sender Emails.

41

Don't open unknown sender attachments.

42

Tag unwanted Emails as spam.

CHAPTER FIVE

STORAGE & PRESERVATION & BACKUP

43

Never keep multiple copies of sensitive information (hard copies or soft copies).

44

Shred immediately unrequired sensitive information multiple copies or else if it may be misused by wrong hands.

45

Store & safeguard sensitive information at right place with right person

46

Keep private legal documents in a safe. Life of legal documents such as assets is prolonged. Preservation and safe custody is very critical as they pass through multi generations.

47

Schedule your deliveries to prevent return or theft when you are unavailable

48

Other than sensitive information for rest of the documents scan & organise online handy & secure way at your disposal with your authentication for retrieval.

49

Hide private documents vault key at a secret place. No one other than you should be able to access those documents.

50

Configure cloud storage settings to restrict sharing access to others

51

Always keep backup of personal information.

52

Backup & store personal information and sensitive data encrypted and password protected in USB drive

53

Weekly survey & delete unnecessary data

54

Personal items shouldn't be left unattended

55

Keep your medical records personal. Destroy labels of medications boxes while disposal.

56

Before selling or donating or disposing your device erase all your personal data and restore the device to factory reset.

CHAPTER SIX

PUBLIC BEHAVIOUR

57

Avoid private discussions in public conversations that can threat your privacy. Be mindful and rational while in public. If you have to have private discussion in public just walk out of public place.

58

Use your own charger in public charging stations not charging cord of charging stations

59

Don't share passwords while on public WiFi.

60

Never carryout commercial or financial transactions using public Wi-Fi

61

Never use public WiFi use only your own data when in public

CHAPTER SEVEN

WORK

62

Don't carry private items to work. You are supposed to carry and keep only professional things to work. Private and professional life's are parallel channels. You can't safeguard your private items if you carry them to work.

63

You may have specific personal items related to work. Keep those personal items locked in your custody in office.

64

Don't keep your private information like certificates, bank data etc in work. Private information is only upto self and confidential. You can't share with others. Don't mix private documents with professional documents. You can't ensure integrity of private documents custody in work.

65

In profession, block anyone who don't want to connect professionally

CHAPTER EIGHT

TRAVELLING

66

In a fast travelling world when in public keep close your belongings. You don't know about others. People's attention is drawn towards valuables in public places. Also don't talk anything while in public that reveals your personal details.

67

Modern life style is hectic due to floating population, everybody tries to be cautious as well as vigilant. To win trust and help, always travel with essentials like ID proof, Keys, Wallet, Phone, SOS medicines.

68

Whether you are commuting, tag your belongings with distinct unique colour tag for your luggage for easy identification, storage & retrieval.

CHAPTER NINE

DEVICE SENSORS

69

Keep device camera & microphone disabled when unused. Block camera with tape to prevent unauthorised access by intruders.

70

All modern devices comes with GPS device location feature with inbuilt GPS sensor for tracking location. It is highly advisable to keep disabled your device location for your safety and privacy.

71

Selectively grant access to apps for camera, mic, contacts, location etc.Remove unnessasary & unwanted app access to camera, mic, contacts, location etc

72

Disable through app, webcam and mics when not in use

CHAPTER TEN

SECURITY

73

Whenever there is news of security breach though you are unaffected change account passwords & and also verify your integrity.

74

Investigate unexplained battery drain and distinctive & unexplained user & app file size changes. Keep backup of vital soft documents.

75

Disable notifications on locked screen

76

Use data usage manager for data use. Investigate sudden data usage surge

77

Use standard antivirus and internet security apps and periodic auto scan.

78

Regularly clear cookies, cache and browser history, temporary files generated.

79

Enable firewall & other safety protections in network

80

Be straight forward with family and friends and insist on maintaining your privacy to others.

81

Use standard and reputable Email service provider.

82

Protect and secure your internet connection with VPN.

83

Don't publish address in classifieds.

84

Time to time review & verify for your privacy breach.

85

Use higher standard mobile network in 2G/3G/4G/5G. Higher the standard higher will be security & safety for your device.

86

Use remote wipe apps to delete lost device data

CHAPTER ELEVEN

SOCIAL MEDIA PLATFORMS

87

Educate & guide your kids about configuring privacy settings in social media accounts. Teach your kids what not to do while they are using parents important devices for self.

88

Teach your kids about the differences in between private and public social media postings, limitations and dangers.

89

Allow & limit your social media accounts interactions & networking only up to the persons you known

90

Educate kids (precautions, prevention, caution) about hackers who steal individual identities and make fake profile accounts in social media with same identity as original.

91

Keep messaging apps disabled when not in use.

92

Never synchronise your messaging or social media apps with your device contacts. Your privacy can be invaded.

93

Disable chats setting in online games. gaming shouldn't be played simultaneously while chatting.

94

Keep your social media apps logged out and disabled when not in use

95

Control and limit access to your posts.

96

Don't join in unknown groups.

97

Alert your friends and relatives when your account is hacked.

98

Turn off your location in social media accounts

99

Never share residence address with photos.

100

Restrict profile sharing in professional social media accounts

101

Be specific and control who can message you on social platforms.

102

Never participate in social media surveys.

103

Never remit money through social media.

104

Before communicating validate online dating profiles.

105

Always meet unknown online dating site friend in public

CHAPTER TWELVE

CHILDREN

106

Use parental controls on devices used by kids and children. Educate kids and children about direct and indirect online privacy breaches and threats.

107

Permit kids to use devices only in the presence of family in common room but not alone.

108

Teach your children how to say NO when someone tries to invade their privacy & try to seek their personal details.

109

Pay attention when your children are downloading apps and if necessary guide them according to occurring changes time to time.

110

Educate children about online risks associated with evolution as well as advancement of technology

111

Allow kids to shop online only in the presence of parents.

112

Never permit children to game online with unknown strangers.

113

Watch children when online

114

Wherever available use content filters as required like parental control for kids etc as per requirement.

CHAPTER THIRTEEN

PERSONAL EMERGENCY

115

Create few specific unique code words for your family members. Incase of personal urgency or personal emergency like external threat or medical emergency etc in public or private place they can pronounce by indicating these code words and seek help when they are in a situation of unable to express freely by some compulsion.

CHAPTER FOURTEEN

WIRELESS CONNECTIVITY

116

Don't share your WiFi passwords other than your own family. Ensure adequate strength passwords for your WiFi or Hotspot

117

Disable bluetooth and other wireless communications when not used

CHAPTER FIFTEEN

PSYCHOLOGY

118

Too good to be true is a very harmful tendency in a modern digital and internet world

CHAPTER SIXTEEN

APPLICATIONS

119

Disable or uninstall unused preloaded apps

120

Do atleast research home work on app before installing. Blindly never install apps.

121

Post installation of app, manually configue app settings to suit to your use.

122

Time to time promptly uninstall unused apps.

123

Password protect app containing commercial and sensitive information

124

Disable unnecessary or unwanted apps running in background

125

Delete accounts / apps that no longer used

126

For apps containing commercial or sensitive information prefer paid app over free apps with advertisements.

127

Use advertisement blocking apps and block advertisements embedded along with apps

CHAPTER SEVENTEEN

UPDATING

128

Upgrade operating drivers regularly as and when available

129

Update operating system software as and when update is available from original device supplier.

CHAPTER EIGHTEEN

BROWSING

130

Never visit embarrassing websites to you and others.

131

Be cautious while clicking on results of internet search results. Review before clicking links.

132

Use private internet browser or browsers in private mode while doing important work.

CHAPTER NINETEEN

INTERNET

133

Use secured internet connections like VPNs

134

Disconnect idle devices from internet. Power down unused devices.

CHAPTER TWENTY

COMMERCIAL TRANSACTIONS

135

Set transaction Email and SMS alerts for your credit card, debit card, banking accounts and commercial platforms.

136

While paying or receiving money always mention total in figures and also in words in writing

137

Never pay for things that you have not seen in person.

138

Review and verify your bank statements and other financial transactions regularly. In case of any discrepancy report it to concerned branch or helpline or bank.

CHAPTER TWENTY-ONE

COMMERCIAL PORTALS

139

For financial or commercial transactions use only https websites for security and protection.

CHAPTER TWENTY-TWO

COMMERCIAL PORTALS LOGIN & PASSWORDS

140

Never use auto filled passwords and account details for payment & login while resorting to shopping

141

Don't write your login passwords anywhere especially for banking, financial and commercial websites. Never lend your login passwords to anyone. Remember you are legally responsible for your account not others.

142

Post work always logout of account from commercial websites and other financial websites containing sensitive information

143

Never save payment information in commercial websites. Add payment information only while resorting purchases

144

Freeze inactive financial transaction accounts.

CHAPTER TWENTY-THREE

CREDIT & DEBIT CARDS

145

In case of privacy breach alert credit card companies and bank through Email, phone as well as through written letter immediately.

146

Investigate privacy breach to know extent of its affect of consequences for remedy, recovery and prevention

147

Use latest debit cards and credit cards with advanced embedded security chips to prevent copying

148

Frequently change debit card and credit card PINs. Never write PIN numbers anywhere.

149

Don't share your credit or debit card details with any one or online

CHAPTER TWENTY-FOUR

BANK ACCOUNTS FOR SHOPPING

150

Use dedicated specific bank account for routine, regular normal retail purchases from commercial websites. Maintain bank balance only just to meet the purchase.

151

Use separate specific bank account for other than normal & commercial transactions and for high end value transactions. For high value transactions for two factor authentication OTP(SMS password) use separate mobile phone. Keep this phone number confidential and never share this phone number with any one including your friends and relatives. Don't use data on this mobile number.

CHAPTER TWENTY-FIVE

PERSONAL DETAILS IN COMMERCIAL PORTALS

152

Periodically as well as whenever there is change in your credentials update know your client(KYC) details with financial institutions and banks

153

Update your Email ID and contact phone number with commercial and financial websites whenever there is change

CHAPTER TWENTY-SIX

FINANCIAL INSTRUMENTS

154

Send financial instruments (like checks and DDs etc) only to specific authentic persons and confirm receipt to prevent falling in wrong hands or losing. Also check financial statements and confirm discharge of financial instruments.

155

Keep(number etc) details record of your financial instruments such as cheques and demand drafts and fixed deposits

CHAPTER TWENTY-SEVEN

FINANCIAL ADVISORS

156

Check and verify qualifications, knowledge, experience and integrity of your financial & property advisors.

157

Interact only with selective banks and advisors to avoid confusion.

158

Deal with portfolio managers who are registered with trade regulator

159

Use services of registered brokers

160

In this modern digital financial world understand there is no free advise by anyone, always there may be direct or indirect hidden charges

CHAPTER TWENTY-EIGHT

ASSETS MANAGEMENT

161

If someone is managing your assets on behalf of you. Give only specific power of attorney (POA). However regularly review their transactions

162

Always give instructions to your advisors in writing to avoid confusion.

CHAPTER TWENTY-NINE

CASH HANDLING

163

Legally limited amount only an individual can pay in cash. Hence for higher amounts pay either through demand draft(DD), crossed Cheque or online bank money transfer.

CHAPTER THIRTY

FINANCIAL RECORDS PRESERVATION

164

Safeguard, protect and preserve your legal, insurance, property, financial, bank documents. Normally an individual is required to maintain personal record of past ten years transactions.

CHAPTER THIRTY-ONE

UNAUTHORISED FINANCIAL TRANSACTIONS

165

Temporarily freeze your account by alerting your bank or trader for observed unauthorised transactions. Investigate the cause with the help of banker or trader.

166

All financial errors or irregularities complaints must be filed in writing with concerned branch and a copy to head office. In case of no resolution within specified time limit grievance must be filed with concerned sector government regulator such as central bank OMBUDSMAN for banks

CHAPTER THIRTY-TWO

FINANCIAL INVESTMENTS

167

Start investing as soon as you start earning.

168

Choose your risk appetite for investing depending upon your age, income, obligations, health etc

169

Invest with long-term perspective

170

Keep contingency or emergency fund aside

171

Decide investment risk only based upon your capacity for unexpected loss

172

Certain major expenses like marriage, higher education, health, home purchase, retirement in life are unavoidable. Hence plan continuous stage saving investments with corresponding maturity.

173

Return on your investment must be more than prevalent inflation with reasonable appreciation.

174

Gold is proven time tested safe investment. Gold never betrayed its investors

175

Investments liquidity is more important than returns.

176

Monitor all your investments centralised with the help of portfolio manager, balance risk & return time to time

177

Remember invest is a learned skill it is neither luck nor fate

178

Learn from bad investments

179

Good research efforts prior investment gives good dividends

180

Invest only in what you understand

181

Never risk to trade money you can't lose

182

Keep diversified portfolio. Never put all eggs in one basket.

183

Short list stocks based on market trend and choose few stocks endorsed by technical analysis.

184

Past performance need not be prediction for future performance.

185

While entering any investment or insurance know premature cancellation charges and benefits

186

Know all hidden charges

187

Check credibility track record of with whom you are investing

CHAPTER THIRTY-THREE

BORROWING

188

Borrow only if expenditure is unavoidable, but limit your borrowing up to your repayment capacity

CHAPTER THIRTY-FOUR

INSURANCE

189

Protect you and your family life, income, health, assets, acts of God with suitable insurances. Start insuring at as early age as possible

190

Provide details correctly while insuring or else you may lose claim

191

Have adequate insurance coverage

CHAPTER THIRTY-FIVE

TAXES

192

Never evade taxes. Tax is a legal liability

CHAPTER THIRTY-SIX

EXPENDITURE

193

Pay only for what you need, not for what you want

CHAPTER THIRTY-SEVEN

SOFTWARES

194

Use latest version of applications and software

195

Never compromise safety and security of your device by using pirated software's

196

Be cautious while using shareware software's. Check authenticity of the software source.

197

Regularly renew software licences

Printed by Libri Plureos GmbH in Hamburg,
Germany